Sebastian Amiebenomo

Development of a Generic Fault Checklist for Diesel Generator

Sebastian Amiebenomo

Development of a Generic Fault Checklist for Diesel Generator

LAP LAMBERT Academic Publishing

Impressum / Imprint
Bibliografische Information der Deutschen Nationalbibliothek: Die Deutsche Nationalbibliothek verzeichnet diese Publikation in der Deutschen Nationalbibliografie; detaillierte bibliografische Daten sind im Internet über http://dnb.d-nb.de abrufbar.
Alle in diesem Buch genannten Marken und Produktnamen unterliegen warenzeichen-, marken- oder patentrechtlichem Schutz bzw. sind Warenzeichen oder eingetragene Warenzeichen der jeweiligen Inhaber. Die Wiedergabe von Marken, Produktnamen, Gebrauchsnamen, Handelsnamen, Warenbezeichnungen u.s.w. in diesem Werk berechtigt auch ohne besondere Kennzeichnung nicht zu der Annahme, dass solche Namen im Sinne der Warenzeichen- und Markenschutzgesetzgebung als frei zu betrachten wären und daher von jedermann benutzt werden dürften.

Bibliographic information published by the Deutsche Nationalbibliothek: The Deutsche Nationalbibliothek lists this publication in the Deutsche Nationalbibliografie; detailed bibliographic data are available in the Internet at http://dnb.d-nb.de.
Any brand names and product names mentioned in this book are subject to trademark, brand or patent protection and are trademarks or registered trademarks of their respective holders. The use of brand names, product names, common names, trade names, product descriptions etc. even without a particular marking in this work is in no way to be construed to mean that such names may be regarded as unrestricted in respect of trademark and brand protection legislation and could thus be used by anyone.

Coverbild / Cover image: www.ingimage.com

Verlag / Publisher:
LAP LAMBERT Academic Publishing
ist ein Imprint der / is a trademark of
OmniScriptum GmbH & Co. KG
Heinrich-Böcking-Str. 6-8, 66121 Saarbrücken, Deutschland / Germany
Email: info@lap-publishing.com

Herstellung: siehe letzte Seite /
Printed at: see last page
ISBN: 978-3-659-71610-2

Copyright © 2015 OmniScriptum GmbH & Co. KG
Alle Rechte vorbehalten. / All rights reserved. Saarbrücken 2015

This monograph is a study conducted in the spring of 2008 by Engr. S.O. Amiebenono assisted by Engr. I.I Omorodion and Engr. J.O. Igbinoba with active collaboration of the maintenance department of Okhmu oil Palm Company and Rubber Research Institute of Nigeria.

The research work outline the development of a generic checklist for diesel power generator maintenance for company personal involve in preventive maintenance inspection.

Executive Summary

Now a days, maintenance cost is one of the major operating costs in manufacturing companies. Unexpected breakdowns, replacement and repair expenses from catastrophic failures results in loss of output due to machinery downtime.

Adoption of preventive maintenance significant reduce losses which invariably increase the product quality and cost.

In most developing nations, like Nigeria where power outages is the norm, generating plants on which most firms depend on, for their energy needs, must be kept in continuous operation without mechanical interruption to meet the demands of production.

It becomes imperative that preventive maintenance inspection (PMI) tied to preventive maintenance (PM) interval on power generating plants needs to be enhance.

In the light of the foregoing, this study outlines a methodology of developing a pragmatic generic checklist for fault detection for diesel power generating plants to guide PM personnel on inspection

Foreword

This study was conducted to evaluate the needs assessment of maintenance techniques currently available in Okhomu Oil Palm Company and Rubber research institute of Nigeria with regards to diesel power plant industrial generating plant.

This work was performed by Department of Mechanical Engineering Ambrose Alli University, Ekpoma Nigeria. The Department Principal investigator was by Engr. S.o. Amiebenono assisted by Engr. I.I Omorodion and Engr. J.O. Igbinoba special credit goes to the maintenance department of these firms for making available there personal to be an integral part of these work

ABSTRACT

The objective of every maintenance department is to anticipate repairs and initiate activities that prevent mechanical failures. Manufacturing firms perform various preventive maintenance (PM) activities at specified intervals to achieve this objective. Some intervals evolve from manufacturers' specifications and industry experiences; others are simply handed down over time with no real understanding of where they came from. Effective preventive maintenance inspection (PMIs) also ensures that equipments (diesel power generators) reach their useful service life. This study examines the various problem and faults associated diesel generators in service. The approach to this study consisted of a survey questionnaire, literature review, and interviews conducted with two manufacturing firm as case studies. The survey questionnaire, posted on the maintenance personnel of these firms, produced a total of 48 respondents. A review of the survey responses confirmed that personnel of both firms are equally represented. The sample is random; therefore, it is representative. Model application employing Likert scale and Opinion Discrimination Analysis(ODA) reveal the influential faults that(PMIs) teams should look out for when carrying out PM programs to help reduce the in-service mechanical breakdowns that occur in the operation of diesel power generators.

TABLE OF CONTENTS

	PAGE
LIST OF TABLES	viii
NOMENCLATURE	ix

CHAPTER ONE

1.0	Background of the Study	1
1.1	Statement of the Problem	3
1.2	Aims and Objectives	4
1.3	The Significance of the Study	4
1.4	The Scope and Limitation of the Study	5

CHAPTER TWO

2.0	Literature review	6
2.1	Introduction	6
2.2	Classification of Maintenance	6
2.2.1	Corrective Maintenance or Breakdown Maintenance	7
2.2.2	Predictive Maintenance (PdM) or Condition Based Maintenance (CBM)	7
2.2.3	Reliability Centre Maintenance	8
2.2.4	Preventive/Scheduled/Planned Maintenance	9
2.3	Data Collection Methods	11
2.4	Primary Data	12
2.4.1	Experiments and Survey	12
2.5	Collection of Primary Data	12
2.5.1	Observation Method	12

2.5.2	Interview Method	14
2.5.3	Collection of Data through Questionnaire	15
2.5.4	Collection of Secondary Data	22
2.6	Selection of Appropriate Method for Data Collection	22
2.7	Likert Method	24

CHAPTER THREE

3.0	Research Methodology	26
3.1	Discriminative Power Index (DP)	26
3.2	The Band of the DP Continuum scale	26
3.3	Computation of the Discriminative Power Index (DP)	27
3.4	Generation of Scale Items	27
3.5	Factors responsible for the breakdown of Diesel Power Industrial Generating Plant	28
3.6	Questionnaire Design and Administration	28
3.7	Appeal to Respondents	30
3.8	Choice of Respondents	30

CHAPTER FOUR

4.0	Analysis of result	31
4.1	Obtaining the DP Values	40
4.2	Representation of the DP Profile	40
4.3	Analysis of DP Values	40
4.4	Results and Discussion	41
4.4.1	Lower bandwidth: $0.2 < DP \leq 1.1$	41
4.4.2	Middle bandwidth: $1.1 < DP \leq 2.75$	43

4.4.3	Upper bandwidth: DP> 2.75	44

CHAPTER FIVE

5.0	Conclusion and Recommendations	45
5.1	Conclusion	45
5.2	Recommendation	45
5.3	Reference	46
5.4	Appendices	
	The research questionnaire	**48**

LIST OF TABLE

Table 3.1	Variables responsible for Diesel Power Industrial Generating Plant Breakdown	28
Table 4.1	Data Matrix of Respondents Score	31
Table 4.2	Ranked Data Matrix of some key factors affecting the breakdown of Diesel Power Industrial Generator	33
Table 4.3	Ranked Data Matrix of some key factors affecting the breakdown of Diesel Power Industrial Generator	34
Table 4.4	Ranked Data Matrix of some key factors affecting the breakdown of Diesel Power Industrial Generator	35
Table 4.5	Ranked Data Matrix of some key factors affecting the breakdown of Diesel Power Industrial Generator	36
Table 4.6	First Quarter Average Score $(AVS)_1$	38
Table 4.7	Fourth Quarter Average Score $(AVS)_4$	39
Table 4.8	Lower bandwidth of DP values of variables that lead to breakdown of Diesel Power Generators	41
Table 4.9	Middle bandwidth of DP values of variables that lead to breakdown of Diesel Power Generators	43
Table 4.10	Upper bandwidth of DP values of variables that lead to breakdown of Diesel Power Generators	44

NOMENCLATURE

PM = Preventive Maintenance

PMI = Preventive Maintenance Inspection

DP = Discriminative Power Index

AVS = Average Score

WT = Weighted total

CHAPTER ONE

INTRODUCTION

1.0 Background of the study

The replacement, repair or maintenance of a machine as it deteriorates over time has been studied in many contexts (Rusell, Meller and David, 1996). It is commonly agreed nowadays that preventive maintenance programme can be very successful in improving equipment reliability while minimizing maintenance-related costs (Sofia, P., and George, T, 2006,). By preventive maintenance (PM) we mean all actions carried out to prevent or pre-empt a failure. These actions can take two forms. Firstly, there are inspection based or non-intrusive techniques. These will encompass the relatively complex inspection such as vibration monitoring but should also include the basic techniques: look, listen, touch and smell. Often these basic techniques are ignored but should and can form the basis for detecting and preventing failures. The preventive maintenance system should provide the inspector with details on where the work should be done, when the work should be done. To enable the inspector's time to be used effectively the work list should be output in, for example a route order.

Secondly, preventive maintenance should prompt for services and overhauls on a calendar basis, on hours run or amount of product manufactured. Using preventive maintenance it is possible to have a much better idea of the condition of the equipment and consequently it is possible to increase the percentage of work that can be planned. Early detection of faults will decrease both the number and duration of breakdowns.

There are a few study conducted in the preventive maintenance program (PMP) area. Lee and Rosenblatt study the case where preventive maintenance is performed on a deteriorating machine after the machines produces the economical manufacturing quantity [5]. A methodology for the development of PM using the modern approaches of FMEA, root cause analysis, and fault-tree analysis was undertaken by Eti,, Ogaji And Probert,(2006). As the various studies imply, preventive strategy are critical. However, the most important tenet of PM is to ensure customer satisfaction by ensuring that equipment which are critical to preserving taxpayer investment are kept in continuous operation without mechanical interruptions. Effective PM also allows vital equipment like power generating plants, which in Nigeria, provide alternative power source given the epileptic power supply in the country to reach their useful service life. PM can be classified in various ways. For this study, based on the research undertaken by Federal Transit Administration (2010), it is broken down into three fundamental elements; Inspections, Repairs, campaigns and replacements and Overhauls/refurbish.

Each maintenance personnel must know and understand inspections .Inspections are the most common form of PM and typically consist of three separate and distinct functions: Service line inspections, Operator inspections and preventive maintenance inspection(PMIs).

Service line inspections are generally done daily as generators get refuelled and cleaned. The purpose is to check vital fluid levels, to obtain as much information as possible about the well being of the diesel power industrial generating plant. How service line inspections are conducted depends on the level of automation, which to a large extent is unavailable in the country. With a manual approach, inspections are done with little or no electronic assistance. Manual inspection procedures entails, manually checking engine oil, transmission fluid, and coolant levels; visually checking the engine compartment for leaks, listening for unusual sounds and other abnormalities.

Operator Inspections also provide an excellent daily inspection opportunity, especially when assigned to the same power generating plant where they can become more sensitive to abnormal conditions.

Preventive maintenance inspections (PMIs) are an essential PM element. Maintenance personnel are brought in at established intervals for various inspections and service work on equipments. The intent is to identify and correct problems on a systematic basis before they become more serious. Early detection allows agencies to plan and prioritize repair schedules, order needed parts, and accordingly plan staff allocation. The alternative is addressing failures when they occur in service, resulting in service delays and operational inconveniences. In more extreme cases, undetected and neglected equipment defects can lead to injury. Inspection Checklists and inspection reports provide a useful tool to help guide technicians through each PM activity. The checklist documents that activities did take place on certain equipment at a given interval, which is helpful in satisfying inspection recordkeeping requirements. The inspection report requires operators to list any defect or deficiency that would affect the safety of the equipment or result in its mechanical breakdown.

Having enumerated on the various distinct functions of inspections, as well as the various tools employed to carry-out PM tasks, the focus of this study is essentially on improving on the role of inspections in PM with particular emphasis on PMIs. In pursuit of this aim, this research seeks to identify the key variables that are associated with breakdown of diesel power industrial generating plant, which can form the basis for the development of a generic fault checklist tool for maintenance personnel involved in PMIs.

1.1 STATEMENT OF THE PROBLEM

The objective of every maintenance department is to anticipate repairs and initiate activities that prevent mechanical failures. Preventive maintenance (PM) is a series of planned actions where machinery breakdown are anticipated, and where the overriding intent of these actions is improving the level of reliability of components and systems availability. PMIs are an essential PM element. A crucial factor in establishing PMI intervals is the need to change oil and provide lubrication as specified by the equipment manufacturers. The oil change interval also provides an excellent opportunity to inspect other critical areas and take corrective action based on identified defects. Checklists provide a useful tool to help guide technicians through each PMI activity. The development of a generic inspection checklist tool becomes imperative in improving fault identification during PMI inspection.

1.2 Aim and Objectives

Diesel generating plants have an important role in power plants as well as in industries and commercial installations to meet continuous and emergency standby power requirements for day to- day use. Based on these underlying fact the research work entails the development of a generic checklist for defect identification during preventive maintenance inspection. In the pursuit of these goal the following objectives are employed.

(i)The use of survey questionnaire, literature review, and interviews.

(ii) Model application employing Likert scale and Opinion Discrimination Analysis (ODA)

1.3 The Significance of the Study.

In Nigeria electric power utilities have always employed traditional maintenance approaches mostly consisted of pre-defined activities carried out at regular intervals (scheduled maintenance). However, such a maintenance policy may be quite inefficient: it may be overly costly (in the long run), and may not extend component lifetime as much as possible. It is now increasingly realised that achieving high-quality maintenance requires prevention at source and a focus on identifying and eliminating the cause of equipment deterioration rather than the more traditional approach of either letting the equipment fail before repairing it or "fire fighting" in the case of an emergency. The PMI approach is an integral part of preventive maintenance (PM) and the focus of PMI is the identification and eventual elimination of equipment defects. The gains arising from such inspections would essentially be improved if

PMI teams assigned to diesel power plants, are equip with fault checklist, which is a major thrust of these study.

1.4 The Scope and Limitation of the Study

The research work is centred on preventive maintenance activity carried out on a diesel power industrial generating plant. This study was able to highlighted twenty eight defects which may be responsible for diesel generator failures, althought this may not be sufficient enough to represent all salient faults associated with power generating plants of this type.

CHAPTER TWO

LITERATURE REVIEW

2.1. Introduction

This chapter has two main sections. The first section presents a complete review on various maintenance techniques. Section 2.3 presents a review of key works that utilize simulation models. In Section 2.4, models that introduce and develop age reduction and improvement factor models are presented. Finally, applications of preventive maintenance and replacement scheduling in manufacturing and production systems, service systems, and power systems are reviewed.

2.2 Classification of Maintenance

What is maintenance and why is it performed? Past and current maintenance practices in both the private and government sectors would imply that maintenance is the actions associated with equipment repair after it is broken. The dictionary defines maintenance as follows: "the work of keeping something in proper condition; upkeep." This would imply that maintenance should be actions taken to prevent a device or component from failing or to repair normal equipment degradation experienced with the operation of the device to keep it in proper working order. Unfortunately, data obtained in many studies over the past decade indicates that most private and government facilities do not expend the necessary resources to maintain equipment in proper working order. Rather, they wait for equipment failure to occur and then take whatever actions are necessary to repair or replace the equipment. Nothing lasts forever and all equipment has associated with it some predefined life expectancy or operational life.

The design life of most equipment requires periodic maintenance. Belts need adjustment, alignment needs to be maintained, proper lubrication on rotating equipment is required, and so on. In some cases, certain components need replacement, (e.g., a wheel bearing on a motor vehicle) to ensure the main piece of equipment (in this case a car) last for its design life. Anytime we fail to perform maintenance activities intended by the equipment's designer, we shorten the operating life of the equipment. In addition to waiting for a piece of equipment to fail (reactive maintenance), we can utilize preventive maintenance, predictive maintenance, corrective maintenance and reliability centered maintenance.

2.2.1 Corrective maintenance or Breakdown maintenance

The unscheduled maintenance or repair to return items/equipment to a defined state and carried out because maintenance persons or users perceived deficiencies or failures (Dhillon, 2002). Maintenance carried out after recognizing the fault and to restore the item in proper working condition again, so that it performs the required function. The general philosophy is "If it isn't broken, don't fix it". The shop floor where this kind of maintenance technique is implemented does not spend any money until or unless the system is down. Hence it can lead to high maintenance costs, unanticipated downtime and secondary damage to the machine, catastrophic consequences, unnecessary spares and lack of control. The only advantage is it does not result in unnecessary maintenance, which can be a part of other maintenance management techniques

2.2.2 Predictive maintenance (PdM) or Condition based maintenance (CBM)

The use of modern measurement and signal processing methods to accurately diagnose item/equipment condition during operation. B.S. Dhillon, Ph.D. (2002). The maintenance is initiated when indicators show the sign of faults in the incipient stages. In simple words, the main criterion is to maintain the right equipment at the right time. The practice of CBM is done by acquiring and analyzing the real time data, so that maintenance activities and resources can be prioritized/ optimized accordingly. (Morales, 2002) The common principle of Condition based maintenance is that timely monitoring of the practical mechanical condition, operational efficiency, and few other indicators related to operating condition of machine-trains and process systems will provide the data necessary for ensuring the maximum time between repairs and minimize the number and cost of unscheduled interruptions caused by failures.

Condition based maintenance mostly relies on the tell tale signs before failure such as change in vibration level and pattern, Increased temperature of the parts, wear in the surface detected via analysis of lubricant, change in system performance, motor current change, etc. The task is to look for out for such signs using condition monitoring technique so that the risk of failure and maintenance costs, both decreases.

2.2.3 Reliability Centre Maintenance

Reliability Centered Maintenance can be defined as "an approach to maintenance that combines reactive, preventive, predictive, and proactive maintenance practices and strategies to maximize the life that a piece of equipment functions in the required manner." RCM does this at minimal cost. In effect, RCM strives to create the optimal mix of an intuitive approach and a rigorous statistical approach to deciding how to maintain facility equipment.

Reliability centered approach was founded in the 1960s and initially oriented towards aircraft maintenance. It is now only in the past ten years or so that this concept has started coming to the industry. It directs maintenance efforts at those parts and units where reliability is critical. Gabbar *et al.* (2003) present an improved RCM (automated environment) process as integrated with CMMS. The major components of the enhanced RCM process are identified and a prototype as integrated with the various modules of the adopted CMMS is implemented. Wessels (2003) proposes a cost optimized scheduled maintenance interval that uses costs as the constraint and overcomes quantitative complexity by use of computer/software technology. This interval enables an organization to implement a comprehensive RCM program effectively. Eisinger and Rakowsky (2001) discuss a probabilistic approach in the modeling of uncertainties in RCM. They conclude by saying that these uncertainties in the decision making of RCM might be unacceptable in many practical applications, leading to non-optimum maintenance strategies. An alternative approach for some specified uncertainties are also discussed. Hipkin and Cock (2000) discuss implementation of RCM and TPM with respect to TQM and business process re-engineering (BPR) and show as to how maintenance implementation follows the path of other interventions.

2.2.4 Preventive / Scheduled / Planned Maintenance

Preventive maintenance is an action involving inspection, servicing, repairing or replacing physical components of machineries, plant and equipment by following the prescribed schedule. Preventive maintenance (PM) consists of actions that improve the condition of system elements before they fail(Bris, Chatelet, and Yalaoui, 2003). Weibull, a noted resource for predicting component reliability, defines PM as a schedule of planned maintenance aimed at the prevention of breakdowns and failures . The definition continues by stating: that the primary goal of PM is to prevent the failure of equipment before it actually occurs. It is designed to preserve and enhance equipment reliability by replacing worn components before they actually fail . Recent technological advances in tools for inspection and diagnosis have enabled even more accurate and effective equipment maintenance. The ideal PM program would prevent all equipment failure before it occurs. Preventive maintenance (PM) is an important component of a maintenance activity.

Within a maintenance organization it usually accounts for a major proportion of the total maintenance effort. Some of the main objectives of PM are to: enhance capital equipment productive life, reduce critical equipment breakdowns, allow better planning and scheduling of needed maintenance work, minimize production losses due to equipment failures, and promote health and safety of maintenance personnel.

From time to time PM programs in maintenance organizations end up in failure (i.e., they lose upper management support) because their cost is either unjustifiable or they take a significant time to show results. It is emphasized that all PM must be cost effective. According to Dhillon, (2002), PM elements consist of the following,

Inspection: Periodically inspecting materials/items to determine their serviceability by comparing their physical, electrical, mechanical, etc., characteristics (as applicable) to expected standards

Servicing: Cleaning, lubricating, charging, preservation, etc., of items/ materials periodically to prevent the occurrence of incipient failures.

Calibration: Periodically determining the value of characteristics of an item by comparison to a standard; it consists of the comparison of two instruments, one of which is certified standard with known accuracy, to detect and adjust any discrepancy in the accuracy of the material/parameter being compared to the established standard value.

Testing: Periodically testing or checking out to determine serviceability and detect electrical/mechanical-related degradation.

Alignment: Making changes to an item's specified variable elements for the purpose of achieving optimum performance.

Adjustment: Periodically adjusting specified variable elements of material for the purpose of achieving the optimum system performance.

Installation: Periodic replacement of limited-life items or the items experiencing time cycle or wear degradation, to maintain the specified system tolerance.

In the field of preventive maintenance research, literature reveals a dearth of empirical works have been carried –out. The recent ones include a paper by Chelbi and Ait-Kadi (2004) that presents a mathematical model for joint strategy of buffer stock production and PM for a randomly failing production unit operating in an environment where repair and PM durations are random. Bris et al. (2003) have shown the efficiency of an optimization method to minimize the PM cost of series parallel systems based on the time dependent Birnbaum importance factor and using Monte Carlo simulation (applied with programming tool APLAB) and genetic algorithm. Badıa et al. (2002) demonstrate development of a model for minimizing the cost per unit time of inspection and PM through selection of a unique interval. Various simulation tools and mathematical models are attempted in recent past for minimizing the PM cost. Similarly, various models for optimum PM policy determination have been attempted using selection of unique inspection interval, introduction of degradation ratio, etc. However, these models may be useful to maintenance engineers if they are capable of incorporating information about the repair and maintenance strategy, the methods of failure detection, failure mechanisms, etc. that justify reasonableness of assumptions, and the applicability of model in a given system environment that can give greater confidence in estimates based on small numbers of production data. However, not many applications have been published. Successful PM programs depend on the quality of work done by operators, technicians, and service line personnel who carry them out. The survey carried out so far has revealed that little or no studies on preventive maintenance inspection with emphasizes on developing checklists and other guidance tools to help guide technicians through each PM activity has been done.

2.3 Data collection methods

The task of data collection begins after research problem has been defined and research design/plan checked out. While deciding about the method of data collection to be used for the study, the researcher should keep in mind two types of data viz, primary and secondary. The primary data are those which are collected afresh and for the first time, and thus happen to be original in character. The secondary data, on the other hand, are those which have already been collected by someone else and which have already been passed through the statistical process. The methods of collecting primary and secondary data differ since primary data are to be originally collected, while in case of secondary data the nature of data collection work is merely that of compilation. This section is devoted to the review of methods of data collection.

2.4 Primary Data

2.4.1 EXPERIMENTS AND SURVEY

Primary data is collected during the course of doing experiments in an experimental research. An experiment refers to an investigation in which a factor or variable under test is isolated and its effect(s) measured. In an experiment the investigator measures the effects of an experiment which he conducts intentionally.

Research of the descriptive type and surveys, whether sample surveys or census surveys, primary data is obtain either through observation or through direct communication with respondents in one form or another or through personal interviews. Survey refers to the method of securing information concerning a phenomena under study from all or a selected number of respondents of the concerned universe. In a survey, the investigator examines those phenomena which exist in the universe independent of his action(Kothari and Gaurav 2004)

2.5 COLLECTION OF PRIMARY DATA

There are several methods of collecting primary data, particularly in surveys and descriptive researches. Some important methods are described below:

2.5.1 Observation Method

The observation method is the most commonly used method especially in studies relating to behavioural sciences. In a way we all observe things around us, but this sort of observation is not scientific observation. Observation becomes a scientific tool and the method of data collection for the researcher, when it serves a formulated research purpose, is systematically planned and recorded and is subjected to checks and controls on validity and reliability. Under the observation method, the information is sought by way of investigator's own direct observation without asking from the respondent. The main advantage of this method is that subjective bias is eliminated, if observation is done accurately. Secondly, the information obtained under this method relates to what is currently happening; it is not complicated by either the past behavior or future intentions or attitudes. Thirdly, this method is independent of respondents willingness to respond and as such is relatively less demanding of active cooperation on the part of respondents as happens to be the case in the interview or the questionnaire method. This method is particularly suitable in studies which deal with subjects (i.e. respondents) who are not capable of giving verbal reports of their feelings for one reason or the other.

However, observation method has various limitations. Firstly, it is an expensive method. Secondly, the information provided by this method is very limited. Thirdly, sometimes unforeseen factors may interfere with the observational task. At times, the fact that some people are rarely accessible to direct observation creates obstacle for this method to collect data effectively

In case the observation is characterized by a careful definition of the units to be observed, the style of recording the observed information, standardized conditions of observation and the selection of pertinent data of observation, then the observation is called as structured observation. But when observation is to take place without these characteristics in advance, the same is termed as unstructured observation .unstructured observation is consider appropriate in descriptive studies whereas in an exploratory study the observational procedure is mostly likely to be relatively unstructured.

Controlled and uncontrolled observation is another form of classification of observation. If the observation takes place in the natural setting. It may be termed as uncontrolled observation, but when observation takes place according to definite pre-arranged plans, involving experimental procedure the same is then termed controlled observation. In non- controlled observation, no attempt is made to use precision instruments. The major aim of this type of observation is to get a spontaneous picture of life and persons. It has a tendency to supply naturalness and completeness of behavior, allowing sufficient time for observing it. But in controlled observation, we use mechanical (or precision) instrument as aids to accuracy and standardization. Such observation has a tendency to supply formalized data upon which generalization can be built with some degree of assurance. The main pitfall of non controlled observation is that of subjective interpretation. There is also the danger of having the feeling that we know more about the observed phenomena than we actually do. Generally, controlled observation takes place in various experiments that are carried out in a laboratory or under controlled conditions, whereas uncontrolled observation is resorted to in case of exploratory researches.

2.5.2 Interview Method

Interview is a data collection method in which thoughts and opinions is gathered. It is a relatively simple procedure to obtain knowledge about a person's experience, experiences, values and opinions. (Kothari, et al 2004). Depending on the interview structure, it is possible to collect quantitative information but also qualitative. It is customary to divide the interview into three categories, namely; Unstructured, semi-structured and structured interview. Depending on what is sought through the interview is selected to which is believed to be most preferable. Structured interviews are best suited to quantitative studies, while unstructured is more suited to qualitative (Kothari, et al 2004).

What distinguishes an unstructured interview is that interviewer asks open questions which are then discussed freely. The interviewed person may thus be controlled since every reference to the desired area can be controlled. The aim is that unstructured interview is preferable when the interviewer in advance is unsure of the areas applied for, or has less knowledge about the subject (Kothari, et al 2004). Unstructured interviews provide for this reason qualitative data. Additional advantages of an unstructured interview are deepening the ability of the issues that seem important to the respondent. The summary of the interview is complicated, however, rated value and the method is not suitable for larger scales.

In a structured interview, questionnaires are used where the respondent may, either independently or through the predefined response options, answer questions. To make a structured interview it requires good knowledge of the subject and a pre-desirable area to study. A structured interview is quantitative. Finally, in a semi-structured interview a priori structure of the areas that are sought in the interview has been made but the order is less operative and follow-up questions are also possible. Thus, the respondent is easier to be involved in the layout during the interview and answer in a more free way. A semi-structured interview provides both quantitative and qualitative responses .

As described, interviewing as a data collection method has both advantages and disadvantages. Advantages are usually mentioned to be that an interview is a subjective and flexible approach with the possibility of a deeper analysis of what the interviewed person really feels about a certain area. It's easy to ask the person to develop and explain their reasoning. This will also minimize the risk of misinterpretation. A further advantage which may also prove to be disadvantageous is that it is possible to influence the sample. The disadvantages are that the interviewee must be present throughout the operation. This may affect the interviewing person adversely. In addition, in an interview, allow the respondent to adjust their responses along what he thinks the interviewer are looking for, and thus influence the outcome. Interviews may also be seen as self-reporting data, that is, they do not give a far-reaching conclusion of what an audience likes but convey only what the respondent thinks (Kothari, et al 2004).

2.5.3 Collection of Data Through Questionnaire

This method of data collection is quite popular, particularly in case of big enquiries. In this method a questionnaire is sent (usually by post) to the persons concerned with a request to answer the questions and return the questionnaire. A questionnaire consists of a number of questions printed or typed in a definite order on a form or set of forms. The questionnaire is mailed to respondents who are expected to read and understand the questions and write down the reply in the space meant for the purpose in the questionnaire itself. The respondents have to answer the questions on their own

The method of collecting data by mailing the questionnaires to respondents is most extensively employed in various economic and business surveys. The merits claimed on behalf of this method are as follows:

1. There is low cost even when the universe is large and is widely spread geographically.
2. It is free from the bias of the interviewer; answer are in respondent's own words.
3. Respondents have adequate time to give well thought out answers.
4. Respondents, who are not easily approachable, can also be reached conveniently.
5. Large samples can be made use of and thus the results can be made more dependable and reliable.

The main demerits of this system can also be listed here:

1. Low rate of return of the duly filled in questionnaires; bias due to non-response is often indeterminate.
2. It can be used only when respondents are educated and cooperating.
3. The control over questionnaire may be lost once it is sent.
4. There is inbuilt inflexibility because of the difficulty of amending the approach once questionnaires have been dispatched.
5. There is also the possible of ambiguous replies or omission of replies altogether to certain questions; interpretation of omissions is difficult.
6. It is difficult to know whether willing respondents are truly representative.
7. This method is likely to be the slowest of all.

Before using this method, it is always advisable to conduct 'pilot study' (Pilot Survey) for testing the questionnaires. In a big enquiry the significance of pilot survey is felt very much. Pilot survey is infact the replica and rehearsal of the main survey. Such a survey, being conducted by experts brings to the light the weaknesses (if any) of the questionnaires and also of the survey techniques. From the experience gained in this way, improvement can be effected (Converse, et al 2004).

Quite often questionnaire are considered as the heart of a survey operation. Hence it should be very carefully constructed. If it is not properly set up, then the survey is bound to fail. This fact requires us to study the main aspects of a questionnaire viz. the general form questions sequence and questions formulation and wording. Researcher should note the following with regard to these three main aspect of a questionnaire.

- **General Form:** So far as the general form of a questionnaire is concerned, it can either be structured or unstructured questionnaire. Structured questionnaire are those questionnaires in which there are definite concrete and pre- determined questions. The questions are presented with exactly the same wording and in the same order to all respondents. Resorts is taken to this sort of standardization to ensure that all respondents reply to the same set of questions. The form of the question may be either closed (i.e. of the type 'yes' or 'no') or open (i.e. inviting free response) but should be stated in advance and not constructed during questioning. Structured questionnaires may also have fixed alternative questions in which responses of the informants are limited to the stated alternatives. Thus a highly structured questionnaire is one in which all questions and answers are specified and comments in the respondent's own word are held to the minimum. When these characteristics are not present in a questionnaire, it can be termed as unstructured questionnaire, the interviewer is provided with a general guide on the type of information to be obtained but the exact question formulations is largely his own responsibility and the replies are to be taken down in the respondent's own words to the extent possible; in some situations tape recorders may be used to achieve this goal.

Structured questionnaire are simple to administer and relatively inexpensive to analyze. The provision of alternative replies at times, helps to understand the meaning of the question clearly. But such questionnaires have limitations too. For instance, wide range of data and that too in respondent's own words cannot be obtained with structured questionnaires. They are usually considered inappropriate in investigations where the aim happens to be probe for attitudes and reasons for certain actions or feelings. They are equally not suitable when a problem is being first explored and working hypothesis sought. In such situation, unstructured questionnaires may be used effectively. Then on the basis of the results obtained in pretest (testing before final use) operations from the use of unstructured questionnaires, one can construct a structured questionnaire for use in the main study.

- **Question sequence:** In order to make the questionnaire effective and to ensure quality to the replies received, a researcher should pay attention to the question-sequence in preparing the questionnaire. A proper sequence of questions reduces considerably the chances of individual questions being misunderstood. The question-sequence must be clear and smoothly-moving, meaning thereby that the relation of one question to another should be readily apparent to the respondent, with questions because are easiest to answer being put in the beginning. The firs' few questions are particularly important because they are likely to influence. The attitude of the respondent and in seeking his desired operation. The opening questions should be such as to arouse human interest. The following type of questions should generally be avoided as opening questions in a questionnaire:

a. Questions that put too great a strain on the memory or intellect of the respondent;
b. questions of a personal character;
c. questions related to personal wealth, etc.

Following the opening questions, we should have questions that are really vital to the research problem and a connecting thread should run through successive questions. Ideally, the question- sequence should conform to the respondent's way of thinking. Knowing what information is desired, the researcher can rearrange the order of the questions (this is possible in case of unstructured questionnaire) to fit the discussion in each particular case. But in a structured questionnaire the best that can be done is to determine the question-sequence with the help of a Pilot Survey which is likely to produce good rapport with most respondents. Relatively difficult questions must be relegated towards the end so that even if the respondent decides not to answer such questions, considerable information would have already been obtained. Thus, question-sequence should usually go from the general to the more specific and the researcher must always remember that the answer to a given question is a function not only of the question itself, but of all previous questions as well

- **Question formulation and wording:** With regard to this aspect of questionnaire, the researcher should note that each question must be very clear for any sort of misunderstanding can do irreparable harm to a survey. Question should also be impartial in order not to give a biased picture of the true state of affairs. Questions should he constructed with a view to their forming a logical part of a well thought out tabulation plan.

Concerning the form of questions, we can talk about two principal forms, viz., multiple choice question and the open-end question. In the former the respondent selects one of the alternative possible answers put to him, whereas in the latter he has to supply the answer in his own words. The question with only two possible answers (usually 'Yes' or 'No') can be taken as a special case of the multiple choice question, or can be named as a 'closed question.' There are some advantages and disadvantages of each possible form of question. Multiple choice or closed questions have the advantages of easy handling, simple to answer, quick and relatively inexpensive to analyze. They are most amenable to statistical analysis. Sometimes, the provision of alternative replies helps to clarify the meaning of the question. But the main drawback of fixed alternative questions is that of putting answers in people's mouths" i.e., they may force a statement of opinion on an issue about which the respondent does not infact have any opinion. They are not appropriate when the issue under consideration happens to be a complex one and also when the interest of the researcher is in the exploration of a process. In such situations, open-ended questions which are designed to permit a free response from the respondent rather than one limited to certain stated alternatives are considered appropriate. Such questions give the respondent considerable latitude in phrasing a reply. Getting the replies in respondent's own words is, thus, the major advantage of open-ended questions. But one should not forget that, from an analytical point of view, open-ended questions are more difficult to handle, raising problems of interpretation, comparability and interviewer bias.

In practice, one rarely comes across a case when one questionnaire relies on one form of questions alone. The various forms complement each other. As such questions of different forms are ~" included in one single questionnaire. For instance, multiple-choice questions constitute the basis of a structured questionnaire, particularly in a mail survey. But even there, various open-ended questions are generally inserted to provide a more complete picture of the respondent's feelings and attitudes.

Researcher must pay proper attention to the wordings of questions since reliable and meaningful returns depend on it to a large extent. Since words are likely to affect responses, they should be r properly chosen. Simple words, which are familiar to all respondents should be employed. Words with ambiguous meanings must be avoided. Similarly, danger words, catch-words or words with emotional connotations should be avoided, Caution must also be exercised in the use of phrases t which reflect upon the prestige of the respondent. Question wording, in no case, should bias the answer. In fact, question wording and formulation is an art and can only be learnt by practice. (Converse and Presser, 1986 and Fowler, 1995).

- **Essentials of a good questionnaire**: To be successful, questionnaire should be comparatively short and simple i.e., the size of the questionnaire should be kept to the minimum. Questions should proceed in logical sequence moving from easy to more difficult questions. Personal and intimate questions should be left to the end. Technical terms and vague expressions capable of different interpretations should be avoided in a questionnaire. Questions may be dichotomous (yes or no answers), multiple choice (alternative answers listed) or open-ended. The latter type of questions are often difficult to analyse and hence should be avoided in a questionnaire to the extent possible. There should be some control questions in the questionnaire which indicate the reliability of the respondent. Questions affecting the sentiments of respondents should be avoided. Adequate space for answers should be provided in the questionnaire to help editing and tabulation. There should always be provision for indications of uncertainty, e.g., "do not know," "no preference" and so on. Brief directions with regard to filling up the questionnaire should invariably be given in the questionnaire itself. Finally, the physical appearance of the questionnaire affects the cooperation the researcher receives from the recipients and as such an attractive looking questionnaire, particularly in mail surveys, is a plus point for enlisting cooperation. The quality of the paper, along with its colour, must be good so that it may attract the attention of recipients. (Best,.1970)

Nowadays, the usage of internet is quite popular in collecting the data. Questionnaires are sent through e-mails to randomly or purposely selected people. Their e-mail ids can be taken from some other surveys. The trend of paid surveys had also started. There are some websites which pay for filling the questionnaires.

2.5.4 COLLECTION OF SECONDARY DATA

Secondary data means data that are already available i.e., they refer to the data which have already been collected and analysed by someone else. Data may either be published data or unpublished data. Usually published data are available in:,(a) various publications of the central, state are local governments; (b) various publications of foreign governments or of international bodies and their subsidiary organisations: (c) technical and trade journals; (d) books, magazines and newspapers; (e) reports and publications of various associations connected with business and industry, banks, stock exchanges, etc.; (f) reports prepared by research scholars, universities, economists, etc. in-different fields; and (g) public records and statistics, historical documents, and other sources of-published information.

Recently most data are published on websites .The sources of unpublished data are many; they may he found in diaries, letters, unpublished biographies and autobiographies and also may be available with scholars and research workers, trade associations, labour bureaus and other public/private individuals and organisations.

Researcher must be very careful in using secondary data. By way of caution, the researcher, before using secondary data, must see that they possess following characteristics; reliability, suitability and adequacy of data. it is quite risky to use the already available data. Secondary data are not discarded if they are readily available from authentic sources and are also suitable and adequate for in that case it will not be economical to spend time and energy in field surveys for collecting information. At times, there may be wealth of usable information in the already available data which must be used by an intelligent researcher but with due precaution.

2.6 SELECTION OF APPROPRIATE METHOD FOR DATA COLLECTION

Thus, there are various methods of data collection. As such the researcher must judiciously select the method/methods for his own study, keeping in view the following factors(Kothari, et al 2004):

- **Nature, scope and object of enquiry:** This constitutes the most important factor affecting the choice of a particular method. The method selected should be such that it suits the type of enquiry that is to be conducted by the researcher. This factor is also important in deciding whether the data already available (secondary data) are to be used or the data not yet available (primary data) are to be collected.
- **Availability of funds:** Availability of funds for the research project determines to a large extent the method to be used for the collection of data. When funds at the disposal of the researcher are very limited, he will have to select a comparatively cheaper method which may not be as efficient and effective as some other costly method. Finance, in fact, is a big constraint in practice and the researcher has to act within this limitation.
- **Time factor:** Availability of time has also to be taken into account in deciding a particular method of data collection. Some methods take relatively more time, whereas with others the data can be collected in a comparatively shorter duration. The time at the disposal of the researcher, thus, affects the selection of the method by which the data are to be collected.
- **Precision required:** Precision required is yet another important factor to he considered at the time of selecting the method of collection of data.

But one must always remember that each method of data collection has its uses and none is superior in all situations. Thus, the most desirable approach with regard to the selection of the method depends on nature of the particular problem and on the time and resources (money and personnel) available along with the desired degree of accuracy. But, over and above all this, much depends in the ability and experience of the rest-archer. Bowley(1937) remark in this context is very appropriate when he says that "in collection of statistical data common sense is the chief requisite and experience the chief teacher.

2.7 LIKERT METHOD

Likert scales or summated scales are developed by utilizing the items analysis approach wherein a particular item is evaluated on the basis of how well it discriminates between those person whose total score is high and those whose score is low. Those items or statements that best meet this sort of discrimination test are included in the final instrument.

Thus, summated scales consists of a number of statements which express either a favourable or unfavourable attitude towards the given object to which the respondent is asked to react. The respondents indicates his agreement or disagreement with each statement in the instrument. Each response is a given a numerical score, indicating its favourableness or unfavourableness, and the scores are totaled to measure the respondent's attitude. In other words, the overall score represents the respondent's position on the continuum of favourable – unfavourableness towards an issue (Jonson,et al 2008).

Most frequently used summated scales in the study of social attitudes follow the pattern devised by Likert. For this reason they are often referred to as Likert-type scales. In a Likert scale, the respondents is asked to respond to each of the statements in terms of several degrees, usually five degrees (but at times 3 or 7 may also be used) of agreement of disagreement. For example, when asked to express opinion whether one considers his job quite pleasant, the respondents may respond in any one of the following ways (i) strongly agree (ii) agree (iii) undecided (iv) disagree (v) strongly disagree.

These five points constitute the scale. At one extreme of the scale there is strong agreement with the given statement and at the other, strong disagreement and between them lie intermediate points. This illustrated below:

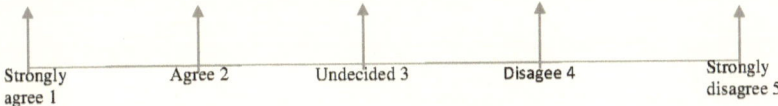

Strongly agree 1   Agree 2   Undecided 3   Disagee 4   Strongly disagree 5

Each point on the scale carries a score. Responses indicating the least favourable degree of job satisfaction is given the least score (say 1) and the most favourable is given the highest score (say 5). These score- values are normally not printed on the instrument but are shown here just to indicate responses. The likert scaling technique, thus assigns a scale value to each of the five response.

Inspite of all the limitations, the Likert-type summated scales are regarded as the most useful in a situation wherein it is possible to compare the respondents the respondent's score with a distribution of scores from some well defined group. They are equally useful when we are concerned with a programme of change or improvement in which case we can use the scales to measure attitudes before and after the programme of change or improvement in order to assess whether our efforts have had the desired effects (Jonson, et al 2008). We can as well correlate scores on the scale to other measures without any concern for the absolute value of what is favourable and what is unfavourable. All this accounts for the popularity of Likert-type scales in social studies relating to measuring of attitudes

CHAPTER THREE

3.0 RESEARCH METHODOLOGY

This chapter presents the methodology applied for generation, processing and analysis of data obtain from field survey which forms the bedrock of this research work.

3.1 DISCRIMINATIVE POWER INDEX (DP)

Discriminative Power Index (DP) is a model employed to distinguish between high and low opinion- ratings. This tends to determine the intensity of the opinions of the respondent on the factors to be evaluated. The higher the DP value, the more controversial the opinions being measured as high DP values indicate divided opinions of the respondents to the issues raised. Lower DP values, indicate popular/a consensus of agreement on factors highlighted. Low DP values, which can possibly be zero indicate con-sensuality/incontrovertibility.

3.2 THE BAND OF THE DP CONTINUUM SCALE

The usual DP continuum scale of opinions has zero (0) as minimum and 5 as maximum thus:

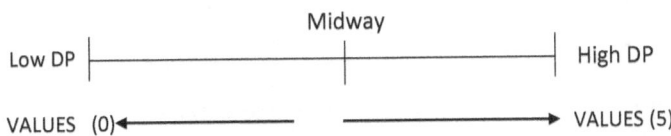

Fig. 3.1: DP CONTINUUM SCALE

DP continuum scale opinion is subdivided into three (3) bandwidths as follows:

(1) **Lower Bandwidth: $0.2 < DP \leq 1.1$**

In this regime, the respondents are in total agreement on the issue considered. in other words there is consensuality

(2) **THE MIDDLE BANDWIDTH: $1.1 < DP \leq 2.75$**

This regime can be described as the average DP. The respondents are fairly in agreement on the subject matter.

(3) **Upper Bandwidth: DP>2.75**

The opinions found with this category shows that respondents are weakly in agreement with what was put forward.

3.3 COMPUTATION OF DISCRIMINATIVE POWER INDEX (DP)

The Discriminative Power is a means to distinguish between high and low opinion ratings. The DP value were obtained by subtracting the 'mean weight or average score of each factor in the 4th Quartile $(AVS)_4$ from its equivalent in the 1st Quartile $(AVS)_1$, that is, $DP = (AVS)_1 - (AVS)_4$

3.4 GENERATION OF SCALES ITEMS:

There are 28 factors (scale items) which are selected to extensively cover the salient aspect of preventive maintenance of diesel power generating plant in Okhomu Oil Palm Company and Rubber Research Institute of Nigeria. These factors are considered sufficiently to reveal the common faults encountered by preventive maintenance inspection teams in the companies mentioned. The choice of these variables which is not brand specific in terms of a given manufacturer of diesel power generator would enhance inspection checklist of maintenance personnel of the companies and other similar organization that are involved in preventive maintenance of power generating plants.

3.5 Factors Responsible For Breakdown Of Diesel Power Industrial Generating Plant

Table 3.1; Variables Responsible for Diesel Power Industrial Generating Plant Breakdown

Vibration	Corrosion
Low Compression Ratio	Overloading
Hardstarting	Dirt in fuel
Loss of Power	Low Volatility of Fuel
Excessive Fuel Quantities	Wrong Oil Specification
Over heating	Over filling 'or oil over gauged,
Excessive Smoke	Engine Deposit
Erratic running	Excessive ware and noise operation
Misfiring'	Lacquering
Damage gasket	Battery leakage (self-discharged)
Low Fuel Pressure	Oil Shortage
Noisy alternator	Open Circuit fault
Short circuit and ground faults	Excessive Current
High rate of combustion	Low voltage (due to worm belt drive)

3.6 Questionnaire Design and Administration

Consequent upon the factors in table 3.1 a questionnaire comprising 28 questions to cover all the factors was developed. The questions carefully constructed to agitate the curiosity of the respondents in order to cause them to respond positively to each of the questions.

The questions were constructed with each having five optional answers of ; (i)Strongly agree (ii)Agree (iii)Undecided (iv)Disagree (v)Strongly Disagree

The respondents were required to tick any one of these options for each of the questions.

Each of the options is given a weighted value for analytical purposes according to Rensis Likert's scales as follows:

Strongly Agree   -5 points

Agree   -4 points

Undecided   -3 points

Disagree   -2 points

Strongly Disagree - 1 points

Procedure: The procedure for developing a Likert –type scale is as follows:

(i) As a first step, the researcher collects a large number of statements which are relevant to the attitude being studied and each of the statements expresses definite favourableness or unfavourableness to a particular point of view or the attitude and that the number of favourable and unfavourable statements is approximately equal.

(ii) After the statements have been gathered, a trial test should be administered to a number of subjects. In other words, a small group of people from those who are going to be studied finally, are asked to indicate their response to each statement by checking one of the categories of agreement or disagreement using a five point scale as stated above.

(iii) The response to various statements are scored in a way that a response indicative of the most favourable attitude is given the highest score of 5 and what with the most unfavourable attitude is given the lowest score, say of 1.

(iv) Then the total score of each respondents is obtained by adding his scores that he received for separate statements.

(v) The next step is to array these scores and find out those statements which have a high discriminatory power. For this purpose, the researcher may select some part of the highest and the lowest total scores, say the top 25 per cent and the bottom 25 per cent. These two extreme groups are interpreted to represent the most favourable and the least favourable attitudes and are used as criterion groups by which to evaluate individual statements. This way we determine which statements consistently correlate with low favourability and which with high favorability.

(vi) Only those statements that correlate with the total test should be retained in the final instrument and all others must be discarded from it.

3.7 APPEALS TO RESPONDENTS

This questionnaire were distributed and the co-operation of the respondents solicited in terms of expressing their candid opinions on each of the technical question along with modalities to express their responses. Conditions of anonymity were guaranteed.

3.8 CHOICE OF RESPONDENTS

The sampling techniques of research methodology propounded by Nachmias (1981) were Adopted in this research. In Okhomu Oil Company and Rubber research Institute of Nigeria the total response from the respondents was forty-eight (48).

These respondents which constitute the sample points are the true representatives of the maintenance team in the companies stated above.

The researcher personally handed over the questionnaire to each of the respondents with a special appeal for their assistance. The questionnaires were retrieved within three days of the distribution with appreciation.

CHAPTER FOUR
ANALYSIS OF RESULTS

4.0 Weighted Scores Of Respondents
Step1: Development of data matrix
The scores of all the 48 respondents for the 28 scale items were entered in the data matrix as shown in Table 4.1 shown below.

TABLE 4.1: Data Matrix of Respondents Score

Respondents	1	2	3	4	5	6	7	8	9	10	11	12	13	14	15	16	17	18	19	20	21	22	23	24	25	26	27	28
1	5	5	5	4	4	5	5	5	5	4	5	5	5	5	5	5	4	4	5	4	5	4	4	5	5	5	5	4
2	4	2	4	3	4	2	5	4	1	5	4	5	4	5	4	4	1	3	5	4	5	5	1	4	3	1	5	2
3	4	2	5	2	3	4	4	4	3	4	5	4	4	5	3	4	2	3	4	4	3	4	3	4	3	2	3	3
4	5	4	4	5	5	5	5	5	5	4	4	5	5	5	5	5	4	5	5	5	5	4	4	5	5	4	5	5
5	4	4	4	2	2	2	5	4	4	2	4	4	5	5	5	5	4	2	5	5	4	4	4	5	4	4	5	3
6	4	2	5	4	5	5	5	5	2	4	4	2	5	4	4	3	1	5	5	5	2	5	2	2	3	4	4	5
7	4	3	5	4	5	5	5	4	4	3	5	3	5	5	4	5	4	4	5	5	2	5	2	2	4	2	3	3
8	5	3	4	4	1	1	2	1	1	1	5	4	4	4	4	4	4	4	4	4	4	4	4	4	4	4	4	4
9	4	5	5	2	4	5	4	3	4	5	5	5	2	4	3	4	3	5	4	5	4	1	3	5	5	4	4	2
10	5	5	5	4	4	5	4	4	4	4	4	5	5	5	5	5	3	5	5	5	5	5	3	5	5	5	5	5
11	5	5	5	4	4	5	5	5	5	4	5	5	5	5	5	5	5	5	5	4	5	4	4	5	5	5	5	4
12	4	2	4	3	4	2	5	4	1	5	4	5	4	5	4	4	1	3	4	4	5	5	1	4	3	1	5	2
13	4	2	5	2	3	4	4	4	3	4	5	4	4	5	3	4	2	3	4	4	3	4	3	4	3	2	3	3
14	5	4	4	5	5	5	5	5	5	4	4	5	5	5	5	5	4	5	5	5	5	4	4	5	5	4	5	5
15	4	4	4	2	2	2	5	4	4	2	4	4	5	5	5	5	4	2	5	5	4	4	4	5	4	4	5	3
16	4	2	5	4	5	5	5	5	2	4	4	2	5	4	4	3	1	5	5	5	2	5	2	2	3	4	4	5
17	4	3	5	4	5	5	5	4	4	3	5	3	5	5	4	5	4	4	5	5	2	5	2	2	4	2	3	3
18	5	3	4	4	1	2	1	1	1	4	4	4	4	4	4	4	4	4	4	4	4	4	4	4	4	4	4	4
19	4	5	5	2	4	5	4	3	4	5	5	5	2	4	3	4	3	5	4	5	4	1	3	5	5	4	4	2
20	5	5	5	4	4	5	5	4	4	5	4	5	5	5	5	5	4	5	5	5	4	5	5	5	4	5	5	4
21	5	5	5	4	4	5	5	5	5	4	5	5	5	5	5	5	4	4	5	5	5	4	4	5	5	5	5	4
22	4	3	4	3	4	5	4	1	5	4	5	4	5	4	4	1	3	5	5	5	5	1	4	1	3	1	5	2

	23	24	25	26	27	28	29	30	31	32	33	34	35	36	37	38	39	40	41	42	43	44	45	46	47	48
1	3	5	3	5	4	4	2	1	5	4	5	2	5	4	3	5	5	3	5	4	5	1	4	5	4	3
2	3	5	5	4	3	4	1	5	4	3	4	1	4	2	3	5	5	5	4	3	4	5	4	4	4	4
3	2	4	4	4	2	4	4	5	4	2	5	4	5	5	2	4	5	4	4	2	4	5	4	5	5	5
4	3	5	4	3	4	4	5	4	3	4	4	5	5	1	3	5	5	4	2	4	4	4	2	4	5	4
5	4	5	5	2	2	4	5	4	1	2	4	5	4	3	4	5	5	5	3	2	4	4	2	4	5	4
6	3	4	2	2	3	5	3	1	1	2	4	3	4	4	3	4	5	2	5	3	4	1	5	4	4	5
7	4	4	5	5	5	4	2	5	5	5	4	2	5	1	4	4	5	5	5	5	5	5	5	4	5	2
8	3	5	4	2	2	5	1	4	2	2	5	2	4	5	3	5	5	4	4	2	4	4	2	5	5	5
9	4	5	5	4	5	4	5	5	4	5	5	5	5	4	5	5	5	5	5	5	5	5	5	5	5	4
10	4	5	5	5	5	4	4	5	5	5	5	4	4	4	4	5	5	5	5	4	5	5	5	5	5	5
11	3	5	2	5	4	4	5	2	4	4	5	4	3	3	5	5	2	1	4	4	2	4	4	4	4	2
12	2	4	4	1	4	4	3	4	1	5	4	3	5	1	2	4	5	4	1	4	4	5	4	5	4	4
13	4	5	5	3	5	4	4	5	3	4	4	4	5	4	4	5	5	3	5	4	5	4	4	4	4	5
14	3	5	2	4	4	5	3	1	4	5	3	5	4	3	5	5	2	4	4	5	1	5	5	5	5	1
15	5	5	5	5	5	4	4	5	5	5	4	4	5	5	5	5	5	4	5	5	4	5	4	5	5	5
16	4	5	5	5	5	4	2	5	5	5	4	1	5	4	4	5	5	5	5	4	5	5	4	5	5	5
17	4	5	4	2	3	4	5	4	1	3	4	5	5	4	5	4	2	3	4	4	3	4	4	3	5	4
18	5	4	4	4	5	5	5	4	5	5	5	5	4	4	4	4	4	4	5	5	4	5	5	4	4	4
19	4	4	2	4	3	4	5	2	4	3	4	5	5	4	4	5	2	4	3	4	2	3	4	5	1	4
20	3	5	4	2	4	4	4	2	4	4	4	5	5	3	4	4	2	4	4	4	4	4	4	4	4	4
21	4	4	4	5	4	1	3	4	5	4	1	3	5	4	4	4	4	5	4	1	4	4	1	4	4	5
22	4	4	4	5	4	1	3	4	5	4	1	3	5	4	4	4	4	5	4	1	4	4	1	4	5	5
23	4	5	5	5	5	2	4	4	5	5	2	4	5	4	5	5	5	5	2	4	5	5	5	5	5	5
24	4	5	5	5	5	4	5	4	5	5	4	5	5	2	4	5	5	5	5	4	4	5	5	5	5	5
25	3	5	2	5	5	4	4	1	5	5	4	5	4	3	5	4	2	5	5	4	1	5	5	4	5	2
26	2	5	2	4	4	4	2	1	4	4	1	4	3	2	5	2	4	4	4	1	4	4	4	4	4	2
27	5	4	4	5	5	3	5	5	4	5	3	5	4	5	4	5	5	3	5	4	5	5	4	5	5	4
28	2	4	4	2	3	3	5	5	2	3	3	5	5	3	2	4	5	4	2	3	3	5	2	3	5	4
29	4	5	4	4	4	5	4	4	4	4	5	4	4	4	4	5	5	4	4	4	5	4	4	4	5	4

Step II: **Ranking of data matrix**

The data matrix of table 4.1 above are ranked in the descending order of magnitude for each of the scale item and divided into four quartiles. This is illustrated in table 4.2

TABLE 4.2: Ranked Data Matrix of Some Key Factors Affecting the Breakdown of Diesel Power Industrial Generator

Respondents	Scale Items																											
														FIRST QUARTER														
	1	2	3	4	5	6	7	8	9	10	11	12	13	14	15	16	17	18	19	20	21	22	23	24	25	26	27	28
1	5	5	5	5	5	5	5	5	5	5	5	5	5	5	5	5	5	5	5	5	5	5	5	5	5	5	5	5
2	5	5	5	5	5	5	5	5	5	5	5	5	5	5	5	5	5	5	5	5	5	5	5	5	5	5	5	5
3	5	5	5	5	5	5	5	5	5	5	5	5	5	5	5	5	5	5	5	5	5	5	5	5	5	5	5	5
4	5	5	5	5	5	5	5	5	5	5	5	5	5	5	5	5	5	5	5	5	5	5	5	5	5	5	5	5
5	5	5	5	5	5	5	5	5	5	5	5	5	5	5	5	5	5	5	5	5	5	5	5	5	5	5	5	5
6	5	5	5	5	5	5	5	5	5	5	5	5	5	5	5	5	5	5	5	5	5	5	5	5	5	5	5	5
7	5	5	4	5	5	5	5	5	5	5	5	5	5	5	5	5	5	5	5	5	5	5	5	5	5	5	5	5
8	5	5	5	4	5	5	5	5	4	5	5	5	5	5	5	5	4	5	5	5	5	5	4	5	5	5	5	5
9	5	5	5	4	5	5	5	5	5	5	5	5	5	5	5	5	4	5	5	5	5	5	4	5	5	5	5	5
10	5	5	5	5	5	5	5	5	5	5	5	5	5	5	5	5	4	5	5	5	5	5	4	5	5	5	5	5
11	5	5	5	4	5	5	5	5	5	5	5	5	5	5	5	5	4	5	5	5	5	5	4	5	5	5	5	5
12	5	5	5	4	5	5	5	5	5	5	5	5	5	5	5	5	4	5	5	5	5	5	4	5	5	5	5	5

TABLE 4.3: Ranked Data Matrix of Some Key Factors Affecting the Breakdown of Diesel Power Industrial Generator

Respondents	Scale Items																											
																			SECOND QUARTER									
	1	2	3	4	5	6	7	8	9	10	11	12	13	14	15	16	17	18	19	20	21	22	23	24	25	26	27	28
13	5	5	5	4	5	5	5	4	4	5	5	5	5	5	5	5	4	5	5	5	5	5	4	5	5	4	5	5
14	5	5	5	4	5	5	5	4	4	4	5	5	5	5	5	5	4	5	5	5	5	5	4	5	5	4	5	5
15	5	5	5	4	5	5	5	4	4	4	5	5	5	5	5	5	4	5	5	5	5	5	4	5	5	4	5	5
16	5	5	5	4	5	5	5	4	4	4	5	5	5	5	5	5	4	4	5	5	5	5	4	5	4	4	5	4
17	5	5	5	4	5	5	5	4	4	4	5	5	5	5	5	5	4	4	5	5	5	5	4	5	4	4	4	4
18	4	4	5	4	5	5	5	4	4	4	5	4	5	5	5	5	4	4	5	5	5	5	4	5	4	4	4	4
19	4	4	5	4	5	5	5	4	4	4	5	4	5	5	4	5	4	4	5	5	5	5	4	5	4	4	4	4
20	4	4	4	4	5	5	5	4	4	4	5	4	5	5	4	5	4	4	5	5	4	5	4	4	4	4	4	4
21	4	4	5	4	5	5	5	4	4	4	5	4	5	5	4	5	4	4	5	5	4	5	4	4	4	4	4	4
22	4	4	5	4	4	5	5	4	4	4	5	4	5	5	4	5	4	4	5	5	4	5	4	4	4	4	4	4
23	4	4	5	4	4	5	5	4	4	4	5	4	5	5	4	5	4	4	5	5	4	5	3	4	4	4	4	4
24	4	3	5	4	4	5	5	4	4	4	5	4	5	5	4	4	4	4	5	5	4	5	3	4	4	4	4	4

TABLE 4.4: Ranked Data Matrix of Some Key Factors Affecting the Breakdown of Diesel Power Industrial Generator

Respondents	Scale Items																											
															THIRD QUARTER													
	1	2	3	4	5	6	7	8	9	10	11	12	13	14	15	16	17	18	19	20	21	22	23	24	25	26	27	28
25	4	3	5	4	4	5	5	4	4	4	4	4	5	5	4	4	4	4	5	5	4	4	3	4	4	4	4	4
26	4	3	5	4	4	5	5	4	4	4	4	4	5	5	4	4	4	4	5	5	4	4	3	4	4	4	4	4
27	4	3	5	4	4	5	5	4	4	4	4	4	5	5	4	4	4	4	5	5	4	4	3	4	4	4	4	4
28	4	3	4	4	4	5	5	4	4	4	4	4	5	5	4	4	4	4	5	5	4	4	3	4	4	4	4	4
29	4	3	4	4	4	5	5	4	4	4	4	4	5	5	4	4	4	4	5	5	4	4	3	4	4	4	4	4
30	4	3	4	4	4	5	5	4	4	4	4	4	5	5	4	4	4	4	5	5	3	4	3	4	4	4	4	4
31	4	3	4	3	4	5	5	4	4	4	4	4	4	5	3	4	4	4	5	4	3	4	3	4	4	4	4	3
32	4	3	4	3	4	5	5	4	4	4	4	4	4	5	3	4	4	4	5	4	3	4	3	4	4	4	4	3
33	4	3	4	3	4	5	5	4	3	3	4	4	4	5	3	4	4	3	5	4	3	4	3	3	4	3	3	3
34	4	3	4	2	4	4	4	4	3	3	4	4	4	5	3	3	4	3	3	4	4	2	3	2	4	3	3	3
35	4	3	4	2	4	4	4	4	3	3	4	4	4	5	3	3	4	3	3	4	4	2	3	2	4	3	3	3
36	4	3	4	2	3	4	4	4	3	3	4	4	4	5	3	3	4	2	3	4	4	2	3	2	3	3	3	3

35

TABLE 4.5: Ranked Data Matrix of Some Key Factors Affecting the Breakdown of Diesel Power Industrial Generator

Respondents	Scale Items — FOURTH QUARTER																											
	1	2	3	4	5	6	7	8	9	10	11	12	13	14	15	16	17	18	19	20	21	22	23	24	25	26	27	28
37	4	2	5	2	2	4	4	4	3	3	4	3	4	5	3	3	4	2	3	4	4	2	3	2	3	3	3	3
38	4	2	5	2	2	4	4	4	3	3	4	3	4	5	3	3	4	2	3	4	4	2	3	2	3	3	3	3
39	4	2	5	2	2	4	4	4	3	3	4	3	4	5	3	3	4	2	3	4	4	2	3	2	3	3	3	3
40	4	2	4	2	2	4	4	3	2	2	4	3	4	4	2	2	4	1	2	4	4	2	3	2	2	3	3	2
41	4	2	4	2	2	4	4	3	2	2	4	3	4	4	2	2	4	1	2	4	4	2	3	2	2	3	3	2
42	4	2	4	2	2	4	4	3	2	2	4	3	4	4	2	2	4	1	2	4	4	2	3	2	2	3	3	2
43	4	2	4	2	2	4	4	1	1	2	4	2	4	4	2	2	4	1	1	4	4	2	2	1	2	3	2	2
44	4	2	4	2	2	2	4	1	1	2	4	2	4	4	1	2	3	1	1	4	4	2	2	1	2	3	2	2
45	4	2	4	2	1	2	2	1	1	1	4	2	4	4	1	1	3	1	1	4	4	2	1	1	2	2	2	2
46	4	2	3	1	1	2	2	1	1	1	4	1	4	4	1	1	3	1	1	4	4	2	1	1	2	2	2	2
47	4	2	3	1	1	1	2	1	1	1	4	1	4	4	1	1	3	1	1	4	4	2	1	1	2	2	2	1
48	4	2	3	1	1	1	2	1	1	1	4	1	3	4	1	1	3	1	1	4	4	2	1	1	1	1	2	1

Step III: Computation of The Average Scores for The 1st (Avs)$_1$ and 4th (Avs)$_4$ Quartiles

The 1st and 4th Quartiles are isolated and worked upon to enable the computation of Discriminative power (DP) values as follows:

The sum of the scores for each of the scale items and their averages 1st and 4th Quartiles are computed and recorded. For example, the average scores for the 1st and 4th quartiles denoted respectively as (AVS)$_1$ and (AVS)$_4$ for factor 1 were computed as follows:

Sum of scores for factor 1 of the 1st quartile is given by:

(WT)$_1$ = Total score x number of respondents in the 1st quartile for factor 1

(WT)$_1$ = Weighted total = 5 x 12 = 60

The weighted mean or the Average score of the 1st Quartile (AVS), for factor1 is given as follows:

Average Score (AVS)$_1$ is given as

$$(AVS)_1 = \frac{WEIGHTED\ TOTAL(WT)_1}{NO\ OF\ RESPONDENTS\ IN\ 1^{ST}\ QUARTILE} = \frac{60}{12} = 5.00$$

Sum of scores for factor 1 of the 4th quartile is given by:

(WT)$_4$ = Total Score x. Number of respondents in the 1st Quartile for factor 1

(WT)$_4$ = 4x12 = 48

$$factor\ 1\ (AVS)_4 = \frac{WEIGHTED\ TOTAL(WT)_1}{NO\ OF\ RESPONDENTS\ IN\ 1^{ST}\ QUARTILE} = \frac{48}{12} = 4.00$$

The WT ,(AVS) and discriminative power(DP) values for all the factors in both 1st and 4th quartiles were similarly computed and presented as shown in table 4.5 and 4.6 respectively as shown overleaf below:

TABLE 4.6 : FIRST QUARTER AVERAGE SCORE – (AVS)1

Scale-Factors

Respondents	1	2	3	4	5	6	7	8	9	10	11	12	13	14	15	16	17	18	19	20	21	22	23	24	25	26	27	28
1	5	5	5	5	5	5	5	5	5	5	5	5	5	5	5	5	5	5	5	5	5	5	5	5	5	5	5	5
2	5	5	5	5	5	5	5	5	5	5	5	5	5	5	5	5	5	5	5	5	5	5	5	5	5	5	5	5
3	5	5	5	5	5	5	5	5	5	5	5	5	5	5	5	5	5	5	5	5	5	5	5	5	5	5	5	5
4	5	5	5	5	5	5	5	5	5	5	5	5	5	5	5	5	5	5	5	5	5	5	5	5	5	5	5	5
5	5	5	5	5	5	5	5	5	5	5	5	5	5	5	5	5	5	5	5	5	5	5	5	5	5	5	5	5
6	5	5	5	5	5	5	5	5	5	5	5	5	5	5	5	5	5	5	5	5	5	5	5	5	5	5	5	5
7	5	5	5	5	5	5	5	5	5	5	5	5	5	5	5	5	5	5	5	5	5	5	5	5	5	5	5	5
8	5	5	5	4	5	5	5	5	4	5	5	5	5	5	5	5	4	5	5	5	5	5	4	5	5	5	5	5
9	5	5	5	4	5	5	5	5	4	5	5	5	5	5	5	5	4	5	5	5	5	5	4	5	5	5	5	5
10	5	5	5	4	5	5	5	5	4	5	5	5	5	5	5	5	4	5	5	5	5	5	4	5	5	5	5	5
11	5	5	5	4	5	5	5	5	4	5	5	5	5	5	5	5	4	5	5	5	5	5	4	5	5	5	5	5
12	5	5	5	4	5	5	5	5	4	5	5	5	5	5	5	5	4	5	5	5	5	5	4	5	5	5	5	5
Total	60	60	60	55	60	60	60	60	56	60	60	60	60	60	60	60	55	60	60	60	60	60	55	60	60	60	60	60
Avs	5	5	5	4.6	5	5	5	5	4.7	5	5	5	5	5	5	5	4.6	5	5	5	5	5	4.6	5	5	5	5	5

TABLE 4.7: FOURTH QUARTER AVERAGE SCORE-(AVS)4

Respondents	Factors																											
	1	2	3	4	5	6	7	8	9	10	11	12	13	14	15	16	17	18	19	20	21	22	23	24	25	26	27	28
37	4	2	5	2	2	4	4	4	3	3	4	3	4	5	3	3	4	2	3	4	4	2	3	2	3	3	3	3
38	4	2	5	2	2	4	4	4	2	3	4	3	4	5	3	3	4	2	3	4	4	2	3	2	3	3	3	3
39	4	2	5	2	2	4	4	3	2	2	4	3	4	5	2	3	4	2	3	4	4	2	3	2	2	3	3	3
40	4	2	4	2	2	4	4	3	2	2	4	3	4	4	2	2	4	1	2	4	4	2	3	2	2	3	3	2
41	4	2	4	2	2	4	4	3	2	2	4	3	4	4	2	2	4	1	2	4	4	2	3	2	2	3	3	2
42	4	2	4	2	2	4	4	3	2	2	4	3	4	4	2	2	4	1	2	4	4	2	3	2	2	3	3	2
43	4	2	4	2	2	2	4	1	1	2	4	2	4	4	2	2	4	1	1	4	4	2	3	1	2	3	2	2
44	4	2	4	2	2	2	2	1	1	2	4	2	4	4	2	1	3	1	1	4	4	2	3	1	2	3	2	2
45	4	2	4	2	1	2	2	1	1	1	4	2	4	4	1	1	3	1	1	4	4	2	3	1	2	2	2	2
46	4	2	3	1	1	2	2	1	1	1	4	2	4	4	1	1	3	1	1	4	4	2	3	1	2	2	2	2
47	4	2	3	1	1	1	2	1	1	1	4	1	4	4	1	1	3	1	1	4	4	2	1	1	2	2	2	1
48	4	2	3	1	1	1	2	1	1	1	4	1	3	4	1	1	3	1	1	4	4	2	1	1	2	1	2	1
Total	48	24	48	21	20	34	38	26	19	22	48	28	47	51	22	22	43	15	21	48	48	24	30	18	26	31	30	25
Avs	4	2	4	1.8	1.7	2.8	3.1	2.1	1.5	1.8	4	2.3	3.9	4.3	1.8	1.8	3.5	1.3	1.8	4	4	2	2.5	1.5	2.2	2.5	2.5	2.2
DP values	1	3	1	2.8	3.3	2.2	1.9	2.9	3.2	3.2	1	2.7	1.1	0.7	3.2	3.2	1.1	3.7	3.2	1	1	3	2.1	3.5	2.8	2.5	2.5	2.8

STEP IV
4.1 OBTAINING THE DP VALUE

The DP Value are obtained only from the 1st and 4th Quarters, therefore, the 2nd and 3rd quarters have no bearing in the DP Computations and are therefore ignore or discarded. Considering the DP for factor 1

$DP = (AVS)_1 - (AVS)_4$

$DP = 5.00 - 4.00 = 1.00$

This model seeks to determine the DP values which are require in determining the most influential variables responsible for diesel generator breakdown.

4.2 Representation of the DP Profile
The values of DP obtain from table 4.6 are highlighted on a Bar chart as shown in figure 4.1 below

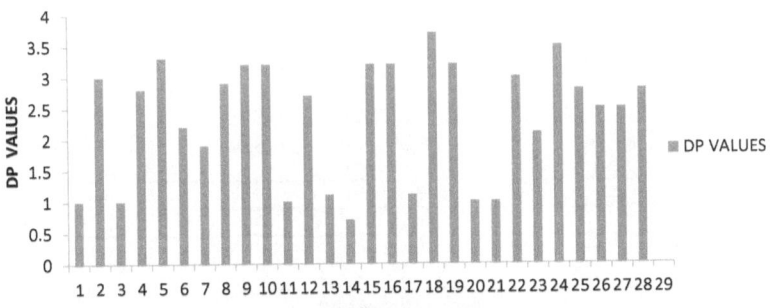

Figure 4.1: DP Values for Twenty Eight Generator Maintainence Variables

4.3 Analysis of DP Values

The DP values tend to measure the intensity level of the variable measured. The higher the DP, the more controversial the opinion measured. Conversely the lower the DP, the more popular the scale item (opinion) measured. A high DP is an index of divided opinion, whereas a very low DP, which may possibly be zero, indicates incontrovertibility-the opinion is widely accepted in an opinion poll. However, it is possible to have an opinion that is mid-way. The attitude of such respondents can be variously described as damped pendulum or middle of the road. They are in the region of "it all depend" or "I am indifferent".

Opinions of this type by respondents are not reliable and not much information can be realised from such respondents. The indifference of this group can only be explained by the clear rating of hypothetical statement. It is a matter of attitude. According, we would make inferences from three main groups. The value of DP for each scale item determines the opinion the respondents chose to fall into. The three bandwidths in the attitude continuum are (i) Conclusive bandwidth ;(ii) Middle of the road bandwidth ;(iii) The inconclusive bandwidth.

4.4 Results and Discussion

The DP values were plotted as bar chart and interpretations rendered. The DP-values are segmented as follows:

4.4.1 Lower bandwidth: $0.2 < DP \leq 1.1$

In this regime, the respondents are in total agreement on the issue considered. In other words there is consensuality.

Table 4.8: Lower bandwidth of DP values of variables that lead to breakdown of diesel power generators

Scale item	Description	DP values
14	Dirt in fuel	0.7
1	Vibration	1
3	Hard starting	1
11	Low fuel pressure	1
20	Lacquering	1
21	battery leakage	1
13	Overloading	1.1
17	Engine over full with engine oil	1.1

FACTOR 14: DIRT IN FUEL:

The low value of DP of 0.7 clearly indicate that Dirt in fuel is the most significance fault preventive maintenance team need to look out for during PM inspection.it results in blockage of the fuel line. It also leads to insufficient supply of fuel within the fuel system. This can be avoided by cleaning the fuel tank thoroughly before fuel is employed for usage.

FACTOR 1: VIBRATION:

Result: DP = 1.00, based on this results, it is reveal that industrial power plants breakdown during operation are majorly due to vibration. It is therefore advisable that thorough monitoring should be carry out by the maintenance team, in order to avoid plant in these companies

FACTOR 3: HARD STARTING FAULT

Results: DP = 1.00, the consideration of the DP value (1.00) reveal that Hard starting occurs a result of wrong injection timing. It is the opinion of the researcher that maintenance personnel should give adequate attention to such fault quickly as possible because of their harmful effect equipment availability.

FACTOR 11: LOW FUEL PRESSURE

Results: DP = 1.00, the DP value prove that low fuel pressure is caused as a result of faulty feed pump and leakage in feed pump suction. The maintenance staff of the organization should ensure that adequate replacement is done to the faulty equipment.

FACTOR 20: LACQUERING
Result: DP = 1.00, based on this result, lacquering is a critical factor that actually lead to engine breakdown as a result of incomplete combustion of the fuel. This fault can be rectified by overhauling the engine with the trained maintenance personnel's to restored the system to an acceptable condition.

FACTOR 21: BATTERY LEAKAGE (SELF DISCHARGE)

Results: DP = 1.00, It can be deduce from the low DP (1.00) that battery leakage (self-discharged) is often the most recurring fault in a generating plant. The OOPC and RRIN should ensure that replacement should be done to an acceptable condition inspection.

FACTOR 13: OVERLOADING

Results: DP = 1.10, It can be inferred from the low DP (1.10) that overloading the motor would cause the engine to malfunction and reduce its life span.

FACTOR 17: OVER FILLING OR OIL OVER GAUGED

Result: DP = 1.1 Considering the low DP value, it is clear evidence that Over gauged may lead to engine malfunctioning. This problem can be avoided by gauging the oil accurately in the system (not below & above the gauge level).

4.4.2 Middle bandwidth: $1.1 < DP \leq 2.75$

In this group, the respondents are fairly in agreement on the subject matter.

Table 4.9: Middle bandwidth of DP values of variables that lead to breakdown of diesel power generators

Scale item	Description	DP values
7	Excessive smoke	1.9
6	Over heating	2.2
23	Noisy alternator	2.1
26	Excessive current	2.5
27	Low voltage	2.5
12	Corrosion	2.7

4.4.3 Upper bandwidth: DP>2.75

In this category, the respondents are weakly in agreement with what was put forth.

Table4.10: Upper bandwidth of DP values of variables that lead to breakdown of diesel generators

Scale item	Description	DP values
4	Loss of power	2.8
28	High rate of combustion	2.8
25	Open circuit fault	2.8
8	Erratic running	2.9
2	Low compression ratio	3.0
22	Oil shortage	3.0
9	Misfiring	3.2
10	Damage gasket	3.2
15	Low volatility of fuel	3.2
16	Wrong oil specification	3.2
19	Excessive wear and noise operation	3.2
5	Excessive fuel quantity	3.3
24	Short circuit and ground faults	3.5
18	Engine deposits	3.7

A multidimensional survey of common mechanical faults that influence the eventual breakdown of diesel power generators in industries was conducted and Rensis Likert's 5-point attitudinal scale was used in dimensioning the respondents' responses, which were captured in a data matrix that was subsequently analysed with opinion discrimination analytical tool to obtain the discriminative powers (DP). The DP-values obtained have enabled us to appreciate the main variables to blame for generator dysfunction. In particular their relative influence is indicative of them magnitude of the DP-values. This is also a technical maintenance guide to action. DP-values established have been helpful in dealing with generator dysfunction common in small and medium scale enterprises especially in developing countries where power generation are stymied by economic depression. Thus the use of these generic fault checklist tools as an aid in preventive maintenance inspection(PMI) would likely engender economic growth and development in the industrial sector.

CHAPTER FIVE

CONCLUSION AND RECOMMENDATION

CONCLUSION

The work has been able to evaluate each of the variables highlighted in the questionnaire distributed to the maintenance department of the various companies. From the analysis of the result, it's been able to enumerate the most important faults (lower bandwidth) responsible for the failure of diesel power engine of which the generating plant is of particularly interest ,which major energy provider in Nigeria

Several problems and faults work conjointly to affect diesel power generators dysfunction. Investigation made through this study revealed that it takes a substantial amount of insight, ability, and experience to establish service intervals and carry out inspection, repair, and overhaul activities in such a way that the collective actions are effective at preventing minor mechanical issues from resulting in mechanical breakdowns that disrupt production process. The opinion discrimination modelling approach proposed is effective in dealing with the problem of identifying the key faults associated with diesel generator in service. Providing technicians with PM checklists as propose in these study would be helpful in guiding technicians through the inspection process which helps to properly complete PM inspections, repairs, and other maintenance activities associated with diesel power generators.

RECOMMENDATION

The result from this work can be used as a generic checklist tool for diesel power industrial generator for most maintenance department of companies employing preventing maintenance policy.

REFERENCE

Anderson, D. (2003). Reducing the Cost of Preventive Maintenance. Business Analyst – Maintenance Oniqua EnterpriseAnalytics.Retrieved September 20,2012 from .http://www.plantmaintenance.com/articles/PMCostReduction.pdf

Bandi, R.K., Vaishnavi, V.K. and Turk, D.E. (2003), "Predicting maintenance performance using object-oriented design oriented metrics", IEEE Transactions on Software Engineering,

Best,J.W.(1970).Research in education(2nded) Printice . Hall,INC.,Englewood Cliffs,New Jersey

Bowley A. L(1937)Elements of Statistics,6th ed., London; P.S king and Staples Limited.

Box, G.E.P., Jenkins G.M. and Reinsel G.C., (1994). Time Series Analysis Forecasting and Control. McGraw- Hill Inc., USA

Bris, R., Chatelet, E., and Yalaoui, F. 2003. New Method to Minimize the Preventive Maintenance Cost of Series-Parallel Systems, Reliability Engineering and System Safety .

Chelbi, A. and Ait-Kadi, D. (2004), "Analysis of a production/inventory system with randomly failing production unit submitted to regular preventive maintenance", European Journal of Operational Research, Vol. 156 No. 3, pp. 712-8.

Converse, J.M., and Presser .S (1986). Survey Questions: Handcrafting the Standardized Questionnaire. Sage University Paper Series on Quantitative Applications in the Social Sciences, 07-063. Thousand Oaks, California. Sage Publications. .

Couper, M. P., Judith T. L, Martin E.A., Rothgeb, J. M and Singer..E.(2004). Methods for Testing and evaluating survey questionnaires. Hoboken, New Jersey. John Wiley and Sons, Inc.

Dhillon, B.S. (2002) Engineering Maintenance, A Modern Approach Washington, D.C; CRC Press

Eisinger, S. and Rakowsky, U.K. (2001), "Modeling of uncertainties in reliability centered maintenance – a probabilistic approach", Reliability Engineering and System Safety, Vol. 71 No. 2, pp. 159-64.

Eti M.C., Ogaji S.O.T., Dan S.D., and Probert, T. (2006). The. Reliability of the AFAM Electric Power Generating Station ,Nigeria. Retrieved October 21,2012 from http://www.cranfield.ac.uk/imrc/FinalReport/Publications/imrcpublications.pdf

Fowler, F.J. (1995) Improving Survey Questions: Design and Evaluation. Applied Social Research Methods Series, 38. Thousand Oaks, California. Sage Publications. .

Gabbar, H.A., Yamashita, H., Suzuki, K. and Shimada, Y. (2003), "Computer-aided RCM-based plant maintenance management system", Robotics and Computer-integrated
Hipkin, I.B. and Cock, C.D. (2000), "TQM and BPR: lessons for maintenance management",

Igboanugo, A.C. and Nwobi-Okoye C.C., (2011).Production process capability measurement and quality control using transfer functions. J. Nigerian Assoc. Math. Phys., 19: 453-464.

Jonson, B. & Christensen (2008).(3rd ed) Quantitative, Qualitative and mixed Research Approach Los Angeles: SAGE Publications.

Kothari C.R., and Gaurav G. ,(2004).Research Methodology; Methods and Techniques.New Delhi,;New Age publication International limited

Lee, H. I., and Rosenblatt, M.J., (1987). Simultaneous Determination of Production cycle and Inspection Schedules in a Production System, Management Science 33/9, pp. 1125-1136.

Morales, D. K. (2002). CBM Policy Memorandum. Washington DC: Deputy under Secretary of Defense for Logistics and Material Readiness.

Rausand, M. (1998), "Reliability centered maintenance", Reliability Engineering and System Safety, Vol. 60 No. 2, pp. 121-32.

Rusell, D. Meller and David S. Kim. (1996) The Impact of Preventive Maintenance on System Cost and Buffer Size, European Journal of Operational Research 95. pp 577-591.
Safety, Vol. 60 No. 2, pp. 121-32.

Sofia, P. and George, T. (2006). Optimal Preventive Maintenance for equipment with two states and general failure time distributions, European Journal of Operational Research. In Press, Corrected Proof. Online 13 June 2006.

Trochim M.K. (2006). Research Methods knowledge Base. Retrieved November 20, 2012 from www.socialresearchmethods.net/kb/scallik.php

Wessels, R.W. (2003), "Cost optimized scheduled maintenance interval for reliability centered maintenance", Proceedings Annual Reliability and Maintainability Symposium IEEE, pp. 412-6.

Worsham W.C. (2005). Is Preventive Maintenance Necessary? Focus on reliability. Column at Maintenance Resources.com. Retrieved August 25, 2012 from http://www.reliability.com/articles/reliability_articles.htm

Department of Mechanical Engineering,
Faculty of Engineering,
Ambrose Alli University,
P.M.B. 14 Ekpoma
Edo State.

Dear Respondent

ACADEMIC RESEARCH QUESTIONNAIRE

The questionnaire you are about to read comprises of 28 questions (scale items) dealing with some of the variables affecting the breakdown of diesel power industrial generating plant. Your view on each of these questions is highly solicited. There are 5 response options and you are to tick which one of the option best represent your view on the issue raised. The pooled views of all respondents consulted would be useful in improving the existing maintenance techniques for Industrial generating power plant.

The view expressed will be treated with utmost confidentiality and anonymity in line with academic research etiquette.

Please comply as requested.

Yours Sincerely

Engr. S.O Amiebenono

QUESTIONNAIRES

1. In a generating power plant, vibration occurring as result of rubbing of unbalanced components is a major caused of power plant breakdown. 1. Strongly Agree ( ) 2. Agree ( ) 3. Undecided ( ) 4. Disagree ( ) 5. Strongly Disagree ( ).

2. Low compression ratio in the cylinder is a regular fault in engine generator. 1. Strongly Agree ( ) 2. Agree ( ) 3. Undecided ( ) 4. Disagree ( ) 5. Strongly Disagree ( ).

3. In diesel engine, wrong injection timing will lead to hard starting resulting in major breakdown of engine generator. 1. Strongly Agree ( ) 2. Agree ( ) 3. Undecided ( ) 4. Disagree ( ) 5. Strongly Disagree ( ).

4. Loss in engine power from unfiltered air entering the cylinder is an often recurring fault in power generating plant. 1. Strongly Agree ( ) 2. Agree ( ) 3. Undecided ( ) 4. Disagree ( ) 5. Strongly Disagree ( ).

5. Excessive quantity of fuel leading to incomplete combustion is a minor factor affecting breakdown of diesel power engines 1. Strongly Agree ( ) 2. Agree ( ) 3. Undecided ( ) 4. Disagree ( ) 5. Strongly Disagree ( ).

6. Overheating arising lack of coolant in the cooling system, strongly effects the breakdown of diesel power engines. 1. Strongly Agree ( ) 2. Agree ( ) 3. Undecided ( ) 4. Disagree ( ) 5. Strongly Disagree ( ).

7. Faulty injector pump may increase the density of exhaust smoke, often results in breakdown of diesel power engine malfunctioning. 1. Strongly Agree ( ) 2. Agree ( ) 3. Undecided ( ) 4. Disagree ( ) 5. Strongly Disagree ( ).

8. Erratic running occurring in diesel engine when there is leakage in fuel system, is a common fault in diesel power engines. 1. Strongly Agree ( ) 2. Agree ( ) 3. Undecided ( ) 4. Disagree ( ) 5. Strongly Disagree ( ).

9. Misfiring occurring as a result of air in fuel system be a major factor responsible in diesel engine failure 1. Strongly Agree ( ) 2. Agree ( ) 3. Undecided ( ) 4. Disagree ( ) 5. Strongly Disagree ( ).

10. Damage gasket is a common engine fault that can lead to diesel power engine breakdown 1. Strongly Agree ( ) 2. Agree ( ) 3. Undecided ( ) 4. Disagree ( ) 5. Strongly Disagree ( ).

11. Does low fuel pressure caused as a result of leakage in feed pump suction, faulty feed pump, and faulty relief valve(s), fast-track diesel engine malfunction. 1. Strongly Agree ( ) 2. Agree ( ) 3. Undecided ( ) 4. Disagree ( ) 5. Strongly Disagree ( ).

12. Does the rate of corrosion (wearing of parts due to the presence of water) as a major factor in engine generator breakdown 1. Strongly Agree ( ) 2. Agree ( ) 3. Undecided ( ) 4. Disagree ( ) 5. Strongly Disagree ( ).

13. In your daily operation, does overloading the motor of a diesel engine a principle factor in engine malfunctioning 1. Strongly Agree ( ) 2. Agree ( ) 3. Undecided ( ) 4. Disagree ( ) 5. Strongly Disagree ( ).

14. Does dirt in fuel result in blockage in fuel line and engine malfunction? 1. Strongly Agree ( ) 2. Agree ( ) 3. Undecided ( ) 4. Disagree ( ) 5. Strongly Disagree ( ).

15. Often low volatility of fuel which reduces maximum power output has been adduced for diesel engine failure in generating plants 1. Strongly Agree ( ) 2. Agree ( ) 3. Undecided ( ) 4. Disagree ( ) 5. Strongly Disagree ( ).

16. Wrong specification of oil often leads to faulty engine generator 1. Strongly Agree ( ) 2. Agree ( ) 3. Undecided ( ) 4. Disagree( ) 5. Strongly Disagree ( ).

17. Oil leakage from transmission as a result of over filing may result in engine malfunction 1. Strongly Agree ( ) 2. Agree ( ) 3. Undecided ( ) 4. Disagree ( ) 5. Strongly Disagree ( ).

18. Engine deposit such as carbon, rust, lime which occur on the cylinder walls and head causes pre-ignition of the fuel-air mixture (ignition to the mixture inside the cylinder before the spark occurs 1. Strongly Agree ( ) 2. Agree ( ) 3. Undecided ( ) 4. Disagree ( ) 5. Strongly Disagree ( ).

19. Incorrect valve clearance resulting in noisy operation and excessive wear, most often leads to engine failure. 1. Strongly Agree ( ) 2. Agree ( ) 3. Undecided ( ) 4. Disagree ( ) 5. Strongly Disagree ( ).

20. Lacquering (piston ring sticking) due to incomplete combustion of the fuel, is a critical factor in engine breakdown 1. Strongly Agree ( ) 2. Agree ( ) 3. Undecided ( ) 4. Disagree ( ) 5. Strongly Disagree ( ).

21. Excessive battery leakage (self-discharge) is often the most recurring faculty in your generating plant 1. Strongly Agree ( ) 2. Agree ( ) 3. Undecided ( ) 4. Disagree ( ) 5. Strongly Disagree ( ).

22. Oil shortage in the oil pan due to a low pressure in the system, often effects engine performance 1. Strongly Agree ( ) 2. Agree ( ) 3. Undecided ( ) 4. Disagree ( ) 5. Strongly Disagree ( ).

23. Noisy alternator arising from bad drive belt is a daily fault occurrence in diesel power generating plant. 1. Strongly Agree ( ) 2. Agree ( ) 3. Undecided ( ) 4. Disagree ( ) 5. Strongly Disagree ( ).

24. It has been generally accepted by your maintenance team that short circuits and ground faults as result of damage done during wiring insulation of equipment is primarily responsible for diesel engine failure in generating plant. 1. Strongly Agree ( ) 2. Agree ( ) 3. Undecided ( ) 4. Disagree ( ) 5. Strongly Disagree ( ).

25. It has been suggested that arising from open circuit fault, badly corroded or soiled connections in the system, often responsible for diesel engine failure. 1. Strongly Agree ( ) 2. Agree ( ) 3. Undecided ( ) 4. Disagree ( ) 5. Strongly Disagree ( ).

26. Excessive current flowing in the circuits that lead to electrical damages, fires, is common fault occurrence in diesel engine malfunctioning. 1. Strongly Agree ( ) 2. Agree ( ) 3. Undecided ( ) 4. Disagree ( ) 5. Strongly Disagree ( ).

27. In your own assessments of faults, are you of the opinion that low voltage due to slack belt is very much responsible in for diesel engine failure 1. Strongly Agree ( ) 2. Agree ( ) 3. Undecided ( ) 4. Disagree( ) 5. Strongly Disagree ( ).

28. Do you accept that high rate of combustion very often leads to engine knocking. 1. Strongly Agree ( ) 2. Agree ( ) 3. Undecided ( ) 4. Disagree( ) 5. Strongly Disagree ( ).

I want morebooks!

Buy your books fast and straightforward online - at one of the world's fastest growing online book stores! Environmentally sound due to Print-on-Demand technologies.

Buy your books online at
www.get-morebooks.com

Kaufen Sie Ihre Bücher schnell und unkompliziert online – auf einer der am schnellsten wachsenden Buchhandelsplattformen weltweit!
Dank Print-On-Demand umwelt- und ressourcenschonend produziert.

Bücher schneller online kaufen
www.morebooks.de

OmniScriptum Marketing DEU GmbH
Heinrich-Böcking-Str. 6-8
D - 66121 Saarbrücken
Telefax: +49 681 93 81 567-9

info@omniscriptum.com
www.omniscriptum.com

www.ingramcontent.com/pod-product-compliance
Lightning Source LLC
Chambersburg PA
CBHW031543210526
45464CB00003B/1131